MEMOIRS OF MILITARY WIVES

HIDDEN HEROES

A SHORT STORY ANTHOLOGY

ROBIN MAJOR-OLIPHANT

WITH CONTRIBUTIONS FROM MILITARY SPOUSES AROUND THE WORLD

MEMOIRS OF MILITARY WIVES
ISBN: 978-0-578-63592-7

One2Mpower Publishing LLC. one2mpower@gmail.com www.one2mpower.com Ordering Information: Quantity sales: Special discounts are available on quantity purchases by corporations, associations, and others. For details, contact the publisher at the address above. Orders by U.S. trade bookstores and wholesalers. Please contact One2Mpower Publishing LLC: Printed in the United States of America

DEDICATION

This book is dedicated to every military wife who serves beside her servicemember; as a civilian or in arms. You are not alone. Know that you are appreciated, you have value, and you have purpose. Believe that your purpose is just as significant as the purpose of your husband's because you're still here.

May this work give you hope, comfort, and inspiration to continue to dutifully serve with boldness, confidence, and empower you to thrive in the often-trivial military wife life gracefully.

CONTENTS

ACKNOWLEDGMENTS

Wow! Thank you, God, for showing mercy toward me as I delayed this project year over year, out of fear and uncertainty. Fear of not being a wife long enough and uncertain of whether I knew enough about military wifehood, to carry the torch for a project as dynamic as this. Nonetheless, this project has been pressing upon my heart since I lunged into my dutiful role as a military wife in 2015, but, I deliberately resisted.

While I was fully aware of "what I signed up for" as a military wife, I had no idea what I was doing. I often felt as though I was alone, doing it wrong, and doing it blindly. I needed tangible reassurance that I was not alone. I needed to know that there were women just like me with the same questions and wonders. I needed to know how to be better at gracefully walking out military wife life so that I could help someone else in the future. I also knew that I could not wait around for someone else to fulfill the vision that was placed on the inside of ME. So, again I say thank you, God. For waiting patiently for me to walk in obedience, for giving me everything that I needed, and sending me every one that you have divinely appointed to fulfill this vision you placed within my heart years ago. You get all of the glory.

Military Family I'd be remised if I did not extend a sincere expression of gratitude to you. Thank you for every single solitary smile in passing as I journeyed seemingly alone most days throughout our military communities. Your warm, polite smiles, and kind gestures, often were that little nudge I needed some days to be reminded that I was not walking this thing out alone. Military life is special, maybe sometimes taken for granted, but special just the same. I'm grateful for our one-off conversations, every small act of kindness, and even for every piece of advice through the military social media communities because it all has and will continue to make a difference.

The Magnificent 10 Thank you to the ten ladies who courageously gave the world a glimpse of your private lives. I am grateful that you trusted me and One2Mpower Publishing with your stories, allowing us to bring them to life. Your transparency and vulnerability are not only admirable, your stories will speak to the hearts of many wives for years to come because of it. This is only the beginning, and I'm excited to see how each of you continues to progress and thrive as military wives.

Timisha Taylor During this short period of time of our acquaintance, you have made a greater impact on me; and this project, than you know. Thank you for believing in me, my ideas, and Memoirs of Military Wives when it was just an idea. Your willingness to continue to support this project; when things seemed bleak, speaks volumes about your genuine character, and your compassion for the Mil-Spouse community. Peace, love, and abundance to you and your future business endeavors.

Erin Bowles Ha! Where do I begin?! I almost let the good one get away because of my then standoffish demeanor. You started off as "my husband's supervisor's wife" and progressed on to becoming my after-school sitter, once we relocated to Alabama. Our children's flourishing friendship was the conduit to you becoming one of my most trusted friends, confidants, and arguably my biggest support and cheerleader. Your consistency and unending support of "all things Robin" is valued and appreciated. I love you, boo! Thank you for being you and consistently giving me room to be me.

FOREWORD

Delighted and inspired- that's the feeling I had when I was requested by Robin Major-Oliphant to write the foreword for this book. Since our first introduction, I have admired her. Robin's tenacity to serve, ability to inspire, warmness, as well as her determination to walk in purpose and her passion to push others to their highest potential are admirable. This is why I am not surprised that she decided to create this project specifically designed for military spouses to provide hope, comfort, and the reassurance to know that "we are not alone."

From having thirteen years of prior service in the United States Army, to serving alongside my husband for seven years as a military spouse, I know firsthand the mental, physical and spiritual capacity needed to endure the day to day journey in this role. A journey often filled with lonely nights, video messaging, change of schedules, sending care packages, growing babies, missed life events, purchasing homes, new hobbies, inventing and reinventing yourself...through it all, you need a support system. Having a friend to call or text and ask, "How are you?" "Can I do anything?" is so appreciated, especially when that friend knows firsthand the emotion and experiences you may be going through. Knowing that Robin is there to connect, support, and share experiences is a blessing. The Memoirs of a Military Wives is yet another.

Some of the first words that come to mind as I think of being a military wife are simple, yet complex, and relatable.

Flexible, amazing, awesome, tired, supportive, courageous, resourceful, adaptable, forgotten, independent, dedicated, courageous, sacrifice, invisible, tenacious, committed, thriving, strong…RESILIENT.

The memoirs in this anthology highlight the perfect imperfections of military life by taking you on an incredible journey of love, hope, loneliness, self-discovery, overcoming, and strength. While each memoir captures the commonality of the love for a service member, they also individually embodied an adventure of curveballs that are a valuable part of the process and life of a spouse that makes them unbreakable and inspiring. I find pieces of myself in each story and am thankful for the journey of these women because not only are they supporting me through their transparent testimonies of ups and downs, highs and lows, they are guiding and teaching lessons learned through their ability to overcome and triumph. Gratitude is my ultimate emotion while experiencing this read.

Memoirs of Military Wives, Volume I is merely a snapshot of what we as military wives endure daily. However, every military spouse, regardless of rank, enlistment status, branch of service, race, ethnicity, or geographical location, will make a connection with this book and take something away from it that will directly impact their lives, or the life of another military spouse they know. -June Wilson

Never above you.
Never below you.
Always beside you.
- Walter Winchell

1 MARRIED SINGLE MOTHERHOOD

Unaddressed wounds become subject to infection…
By Timisha Taylor

There I was, just me and my seven-and-a-half-month baby bump on the road from Virginia to California—alone. During this three-day solo journey, I experienced silent moments on the road with my unborn child, not knowing months of loneliness would soon be my reality. I was very excited about our move to the West Coast. Warm weather, living minutes from the beach, palm trees, and all the beautiful scenery to enjoy and explore with my family.

As I pulled up to our empty new house awaiting the movers, I could sense in my husband's eyes, he had something he wanted to tell me. I asked him to go ahead and share the news, and that's when he told me he'd be leaving for the opposite side of the country within the next couple weeks, and I may have to have our first child alone. He said the best he could do was fly in and hope he'd get home in time for the birth, and then he'd have to immediately fly back to the East Coast, leaving me once again alone. The bad news didn't stop there, he then added he would be gone all summer, training in the desert and deploying at the end of the year for a six-month deployment. I immediately regretted our decision to have a child. Everything else in our life was going so right from becoming military officers, to getting our degrees, and traveling the world, but the timing couldn't be more wrong. Here I was in a beautiful place with no family or friends about to embark on "single motherhood" as a wife. How oxymoronic? How is this happening? I did things the "right way!"

Fast forward to March 16, 2017, my husband was in North Carolina, I in California, the contractions kicked in, and I was ready to delivery my first child with no husband in sight! My husband immediately bought a plane ticket and told me he'd be there as soon as he could. That night, the nurse came in and told me the baby could come any minute, and I more than likely would not make it to my husband's arrival.

Noon, March 17, 2017, Dr. Fan came in and said, "Sorry lady, it's time to break the water, and for this baby to come out," the contractions were less than a minute apart and putting the baby under stress. As the medical staff was preparing for me to deliver, my husband walks in in the nick of time to be by my side, and we had a beautiful baby girl. It was almost as if she waited for him. This beautiful, new family bliss would end two days later. My husband had to head back to North Carolina, and there I was alone with a new baby, I had no clue how to care for, with no help in sight to support me.

The nights were long, and the days were rough. I was running a constant mental and fatiguing marathon with not a soul to pass the baton to, and for the first time in my life, I felt as though I had lost control. I was barely functioning, depressed, sleep-deprived, yet always anxious about everything in regard to my new baby. My body and mind never rested and never caught a break.

Three months later, my husband returned home. I recall that day so vividly. It was the first time he had been gone for months, where I wasn't excited or happy to see him return. He leaned in for a kiss, and I walked out the door to just get a quiet moment of liberation that I desperately needed. I no longer cared about him, how his school went, or even how he was doing because I was so lost in suffering and internally screaming for a break of relief. That day was the very first-day bitterness, and resentment showed its face in my marriage.

Growing up as a little girl, I always saw single mothers endure the hardships they experienced and told myself I would do things the "right way." Well, life had a way of laughing at the plans of my perfectly orchestrated order of life. Don't get me wrong. This isn't to compare myself to single moms who are truly single moms because my husband financially provided, but I was so mentally, emotionally, and physically exhausted from carrying the burdens of motherhood alone, all my husband became to me was a paycheck. I've never had such a disconnected length of time and season in my marriage. I was suffering alone while my helpmate was away elsewhere. There were days I didn't want to speak to him; there were days I lashed out, there were days I was bitter—all counterproductive approaches to channel what I was experiencing. I worked hard, built a career for myself in the military, racked degrees, traveled all corners and continents of the world like a nomadic wanderlust and what felt like overnight, I was at the bottom of the totem pole. My kid always came first for me. While my husband's life and career were unchanged, I felt as though I was just the help—helping and serving everyone else but myself, day in and day out.

The love was gone, the romance was gone, my cup was completely empty. I had nothing to give to my marriage. Whenever my husband would come around (briefly), I just wanted to run far, far, away, and escape from it all. I lost the desire to be a wife, I had no choice to be a mom because she didn't ask to be here, but I had to find a way to recover and circle back to myself.

For months and months, I was so sleep-deprived, isolated, and socially disconnected. It was like I had zero identity outside of being a mother. I constantly found myself comparing how great life was before having a child—the awesome career, the businesses, my social life, the camaraderie, my mental health, my happiness with my physical state—and here I was trading that all in for the sake of married, single motherhood. Before becoming a mom, everyone says, "Congrats!" Joy was an emotion I lacked significantly and lacked because I was so consumed in a life that took so much from me.

For the first time in my life, I actually considered therapy just to break out of the rut I was in. No interest in medications, hypnosis, just tools to help me piece things back together. I came across a quote one day that said, "Whatever you are not changing, you are choosing." I knew in order for things to normalize within myself and my marriage, I had to be willing to make a change. It was all very, very hard, and even more shocking to find out how many mothers experience the same thing and just suffer in silence. No amount of combat I've ever experienced compared to the endurance needed for me to break out of the hardships and fatigue I was enduring. You see moms out all the time, and they seem "okay," but are they surviving, or are they thriving? One day I made the decision to work doubly as hard if I had to, to fight through this to get to a better mental state and emotional climate. I also began to question should I even be fighting for the person I was before motherhood? My pre-motherhood days was all I had to hold on to that was a positive reflection of peace, happiness, and a woman who once soared. Many moms told me I would never get the "old me" back, so chasing "her" would be like running on a hamster wheel; instead, embrace the best version I can be in my new season.

Overall, I feel like I would have enjoyed motherhood in a healthier, happier, and more significant way with a present husband or perhaps a husband who just worked nine to five and came home. When the sun sets, this is the life I asked for and said yes to. I wanted to be a mother. I wanted to marry my husband, whose occupation was an active duty Marine, so I had to simply learn to change perspectives and bloom the best I could where I was planted (or stationed). I commend all moms because it's all hard work, no accomplishment has been more challenging that motherhood—it stays, and it sticks. I'm thankful for the assignment of motherhood and find peace in knowing the military will not be our forever, but my family, our legacy, and our journey will live on past our chapter of life in the military.

2 TIPPING THE SCALE

Mil-to-Mil life is hard, but it can work
By An Anonymous Military Wife

My seventeen years in the military have been great, trying, tiring, fantastic, and so many other things. My husband and I are both active duty Air Force, so it puts a strain on our relationship in ways that I was not ready for and had not even considered. We have a daughter who's a blessing; however, in her nine short years of being on this earth, she has made a lot of sacrifices by default due to our careers. My career field is deployable, while my husband's is not. This being the case, while we have been married for eleven years, we've spent the majority of them apart.

My first time departing my family was one month after my daughter's first birthday. I left for Korea for one year for us to move out of Idaho, where we had been for six years. It wasn't the place we wanted to settle down; I was the one eligible to move, so I made the sacrifice. It was a necessary one, I believed. I remember my daughter seeing me through the computer screen for the first time; she leaped out of my husband's arm, and it hurt my feelings. Instantly I felt the guilt of being away from your young child, my husband, my family. At that moment, I wondered if I had made the right choice. Had I abandoned my duty to put my family first? During this same time period, my husband was moving into base housing with our one-year-old baby, and I wasn't there to help. I wondered if my daughter would remember me. Was this what mothers did? What about his job? Was I being selfish? These were the things that crossed my mind often.

Adding more confusion to the equation was the fact that my husband's job also required him to be TDY often, which was difficult for him. We eventually made the decision to allow our daughter to go and stay with my parents for a while. This gave us both peace of mind and assurance because we knew that she was in good hands but watching her grow up via pictures was hard—I cried often. I was blessed to be in Korea with an amazing network of people to keep me lifted. I had a church family, amazing women who helped me through, and with the love of God, I was able to enjoy my time, grow, and work through all those thoughts.

Transitioning back into home life was rough. Again, something that I hadn't much considered but bringing two lives together again after being apart, in two separate worlds is hard. Between deployments, TDYs, military training requirements, and studying for promotions, our household spent more time apart than together. As often as I left my child, I was accompanied by paralyzing guilt. I often felt like I needed to make up for the lost time, and as a result, I tried to overcompensate for many things and failed more often than not. As a wife, I didn't know what I was doing because I spent more time focused on my daughter than my husband. The guilt of leaving my child had far exceeded my guilt for not being present as a wife, and because of that, my focus was primarily on our daughter. When we did have time to spend together as husband and wife, we talked a lot about work and couldn't find a way to talk about anything else. We focused a lot on work or our daughter. We lost ourselves and found it hard to get back on track. I'd be remised if I did not mention that counseling was a try, but an honest fail because we were focused more on our daughter and careers, and less on each other.

Balancing work and family is one of the hardest things to do. When work is going well, home life is off and vice versa—this was true in our household. I found it hard to stay on track and keep balance in my life.

Although I still don't have everything balanced, I was able to figure some things out as a military wife:

-Mil-to-Mil life is hard, but it can work.
-Your family is first in all situations.
-Being imbalanced isn't a bad thing; it's an opportunity to recognize when the scales are tipping too far over.
-Use the tipping point as an opportunity to course-correct, reset expectations, and get back in balance and harmony within yourself, your union, and your family.
-Making time for yourself is essential, making time for your relationship is important, and trying to overcompensate won't change anything.
-Military kids are resilient and are afforded opportunities to travel the world and see things that most kids only read about- take advantage of it.
-There is growth in this military life, no matter what.

3 OVERCOMING RESENTMENT

Bitterness never wins
By Melissa McGill

As I stood by my husband's side, smiling so graciously, displaying a *proud navy wife* demeanor on the outside, my insides were in tears. Here he was receiving yet another award for excellence to the Navy while shining like a star, and there I was standing by his side but feeling like an obstacle in the shadows. I started to sweat as I felt my temperature rise. The more I listened to all the accolades (especially the ones I actually helped him obtain), the angrier I became. It wasn't that my husband didn't recognize me during his moments of success. He always has. What I was feeling was bitterness and resentment of literally being a military spouse.

Before I married my husband, I was an

independent, career-oriented woman. I was the epitome of the song "I N D E P E N D E N T" by Destiny's Child. I had my own house, car, worked two jobs, and I was receiving recognition for my accomplishments. My husband was actually drawn to my independent nature. He would express how he loved my drive and could picture us as a power couple. I could see it too.

Shortly after we were married, the PCS (permanent change of stations) season began. At first, it wasn't too bad. Although I was giving up the closeness of friends and family (which were my support system), I was thankful that I had a portable career that I could take with me. However, at one of our first duty stations, a little bit of reality started to settle in. I was rejected for several jobs. The reasons I was given ranged from being overqualified, underqualified, salary negotiations to contract not accepted…I was in disbelief until another military spouse shared with me that once employers see that you are a military spouse, they don't want to invest in you because they know you will only be in the area short term. Legally, employers can't say that to you, but I know it happens.

I started to become unsettled in my role as a SAHM (stay at home mom) or housewife. My single girlfriends thought I was being ungrateful and crazy altogether, but they were looking from the outside in. It wasn't just about being in corporate America. I was feeling a void. So, I decided to embrace my season and enrolled in school. I figured I could use the time to earn another degree for my future. During that time, it felt like I would take one step towards excelling, but my husband would take two. No, he was not competing with me, but in my mind, that is how it looked.

We relocated to Yokosuka, Japan, in 2011. I was excited to be in such a pleasant country but still feeling a void in my life. The accomplishments of my husband began to pour in like water. He received awards often, became a pillar in the community, and sometimes, people would see me out in the community and refer to me as *CMC's wife*. "No, I have a name, and it's Melissa," is what I wanted to say! Instead, I graciously smiled and nodded my head. Then one day, while having a discussion with my husband, I blurted out of anger, "I didn't have this problem until I became a MILITARY WIFE"! There it was. The bitterness and resentment showed its face, and it was not pretty. I hurt my husband's feelings, and I knew at that moment, I had to redirect my own feelings.

I became heavily involved in the community. I began working full-time as a government employee. I would say yes to every opportunity to lead something or take on a project. However, my plate eventually became too full. I allowed the bitterness and resentment to cause me to be out of balance as I was still a wife and a mother. My husband noticed it too.

We had a very hard but much-needed conversation one night. I expressed exactly how I had been feeling to my husband. I shared how I felt like I had been making sacrifices for our family that went unnoticed. I expressed how even though I was proud of his accomplishments that sometimes I felt lost as if we were not a team. I had seen my husband's career flourish while mine took a back seat and even diminished. I tried to overcompensate and ended up taking on more than I could even balance.

The conversation with my husband was the first step. I had to acknowledge the problem and give it a name. Then, I had to address why I was feeling the way I was. Lastly, I had to find a resolution. To begin with, I called it just what it was: bitterness and resentment. I had to see myself from a different perspective. I am an intelligent, goal-oriented woman. God gave me several gifts and talents, and just because I changed environments didn't mean I lost my gifts and talents. So, I had to challenge myself to utilize them. To address the feelings, I had to remind myself that I was more than a shadow in the background; I mattered too. I could not blame my husband (which I didn't) for excelling. After all, I didn't want a loser for a husband. I knew my husband truly appreciated me because he always recognized me in whatever accomplishment he received. I just had to recognize it. We were a team. I supported him, and he supported me. I had to edit my thoughts because resentment could have torn my marriage apart. I focused on my wants, my desires, and my goals without becoming selfish.

I shifted my thoughts to my gifts and talents. I

knew what I was great at. More importantly, I knew God had a plan for me. I journaled a lot. People loved reading quotes that I wrote. Several people spoke into my life, but because of the mind frame I was in, it took me a while to hear it. "You should write a book," they would say. Finally, I listened. I began my journey to becoming a self-published author. In addition to becoming an author, I revamped the organizations I was involved with in the community. I withdrew from all and started over with the ones I could best serve my purpose in. I no longer took on more than I could actively be productive with. My yes meant yes, and my no meant no. I had to keep a mental balance in my life.

Today, I do not have any bitterness or resentment in my life. Now when my husband is out in the community, he often hears, "You're Mrs. McGill's husband," and he proudly states, "Yes, yes I am"! The journey I took to get to this point was through prayer, self-awareness, acknowledging my feelings, self-reflection, acceptance, vision and planning, execution and balance. God did not create us without a plan for us. I am honored to be a military wife. I am also grateful to have my own identity.

I often speak with other military wives about my journey. The one thing I try to encourage them with is not to lose themselves inside their husband's career. I remind them that their identity is important, as well as their gifts and talents. If they are in a current situation where they cannot find a job outside the home, I suggest that they build their own empire from within the home by utilizing their gifts and talents, obtaining education, especially business management courses, and setting small obtainable goals. Most importantly, I remind them to edit their thoughts. "Idle hands are the devil's workshop" (Proverbs 16:27-29 TLB). Any wife with too much time on her hands may find herself with bitterness and resentment for her husband. Therefore, it is important for her not to lose interest in things that she likes. For example, I was a paralegal when my husband and I married. After a few relocations, I ended up without a place to utilize my paralegal skills. As a result, I enrolled in a criminal justice degree program online to further my paralegal skills. My hands were no longer idle.

Rediscovering myself and overcoming bitterness and resentment has not been an easy task for me. I had some highs and some lows. However, through it all, I discovered not only am I resilient, but I also found some hidden talents that I had long forgotten about. I was so busy being a military wife that I lost sight of being an author and a wife coach.

I encourage each of you to write out your vision for yourself and make it plain. Set small obtainable goals. Further your education (There are so many free resources available to you). Do not take on more than what you can productively handle. You've got this girl!

4 CONNECTED

Loneliness does not discriminate
By Leah Sledge

My story about making connections started long before I became a military wife. I was born in 1970, when segregation and racism were still heavily woven in the fabric of America. As a little black girl with "nappy" hair, loneliness was a way of life that was exacerbated by moving every two to three years. It wasn't until the late 1980s and early 1990s when I became an African American, and my hair became kinky or tightly coiled. I share this part of my story with you because many times, we try so hard to equalize everything that we miss out on the concepts that make us unique within our own right.

Loneliness does not discriminate.

Because of the prejudices that I had to endure growing up, my loneliness stemmed from not loving myself first. I wanted to belong. I wanted to be like the others, but no matter how much I shopped at the thrift store to find the right outfit or sat in the chair to have my hair pressed for special occasions, I was too busy trying to cure my loneliness by being someone else that I missed out on being the best version of me.

Through many issues and tissues, I have been able to break through and grow through those childhood moments of despair and fully realize the gift of my life. For me, overcoming loneliness started by developing a serious relationship with Jesus Christ and understanding that his plans for my life were intentional because my natural loneliness prompted me to look for a friend in him.

Loneliness is a state in the mind. When we have too much time, we think. When situations don't go as planned, we think. When we encounter a new set of circumstances, we think, and unfortunately, most of our thoughts are negative in context, which doesn't help to quell our fears. Instead, our fears began to overwhelm us to the point of missing opportunities for involvement.

In the book of Genesis, there is a passage of scripture where God states it is not good for man to be alone (Genesis 2:18). Most people take the description literally because this is where God put Adam to sleep and formed Eve from his side. One thing to remember is that we were created in God's image. God is not alone: God the Father, God the Son, and God the Holy Spirit. Like the trinity, he has three distinct personalities, but he is one God. He is neither male nor female; He is God. If God, in Genesis, recognized that man should not be alone, and we were created in his image, we also should not be alone. God established the first family unit, and from there, society grew. Not only were families developed, but relationships, in general, have been around since the beginning of time. Our development and maturity are enhanced by having connections and relationships with others.

My first suggestion for making connections would be to encourage you to know exactly who you are. Many times, we "hook up" with people to be a part of the group, and we compromise our morals and standards to fit in. When you know who you are (and Whose you are), there is more opportunity to bond with like-minded individuals because you realize that time is precious, and you don't have time to waste with those who are not going to help you reach your destiny. What do you like to do? Read? Cook? Volunteer in schools? Craft? Hone a skill to keep occupied so that when you find yourself in a time period of aloneness, you can be alone, but you don't have to be lonely.

My second suggestion would be to use your individual skillset to open up opportunities to enjoy the activity with others. If you like to read, what about joining or starting a book club? If you like to cook, there are many places that offer cooking classes, especially on/near military installations. Schools always need adults to volunteer in various capacities and being a part of the booster club or another school-related organization can help to make connections when you are new to the environment.

My next suggestion would be to open your mind to get out of your comfort zone. Develop a growth mindset. In order for a plant to grow and flourish, the seed must die. The seed is the casing that protects the infant stages of the plant. What casing do you have around your heart and mind? Can you come out of your comfort zone/break your casing and do something to show a willingness to get involved? Don't worry if you do not have any of the skills that were suggested in the earlier paragraphs. What can YOU do? What are you willing to try? The only way to grow any capacity is to be willing to be fed and receive what is being shared from others.

Let's go back to the plant analogy. So, the seed has been planted deep in the ground and it has broken the surrounding casing in order to grow into a strong and healthy plant. The growth of a plant doesn't happen overnight. The plant now has to push through the heavy layers of dirt upwards, all the while sending roots downward to anchor the plant for the trying times when it gets above ground. This is you. Knowing who you are and whose you are deals with your root system. You have to know that you know that you know that you are fearfully and wonderfully made (Psalm 139:14). You are going to have fears and doubts to tell you that you are not enough or that you will never fit in, but you must be able to shut down those negative thoughts and focus on the purpose for being—to bring glory to God and to make a difference in the life of someone else. As you push upwards through the layers of "where am I" and "what am I doing", your resolve will be strengthened in your gifts and abilities because we all have something inside of us that makes us who we are—an original and not a copy. Even twins are not 100% the same in every way.

The plant grows upward—towards the light. The light, the sun (Son), provides warmth and needed nutrients to help the plant continue to develop. The sun has "assistance" through the wind and the rain to help the plant bloom/grow. As card-carrying members of the human race, other people are in our lives to "assist" us in growing. Some nurture us. Others might provide us with stability. There are even those who may attack our character and hurt our hearts, but we can still learn from those relationships and continue to grow strong.

As a military wife, I have had many individuals come into my life as well as go out of my life. I have also had it hard as the pastor's wife of an off-base church in an overseas military location. When people come and go from my life (more specifically the church), I over-analyze the situations to the point of stress and worry that I did something wrong for them to leave the church or that it was my fault that they didn't come back. I have now matured to understand that everyone is in PCS mode: at some point in time s/he is going to rotate to another duty station. With this mindset, I have purposed in my heart to be a blessing to them as much as I can while they are in my presence and leave the rest up to God. As different ladies that I developed close relationships left the area, God was faithful to send someone else—not to replace them—but as a reminder that he would never leave me nor forsake me. He sends in just the right people for me to connect with at the right time.

My upbringing during the time of segregation helped to develop and mature me into who I am today. I am a teacher who works with youth, and I have changed from teaching tolerance (25 years ago) to teaching acceptance. Tolerance does not fully engage the heart and mind in our relationships. Acceptance means that I accept the person as an individual (and a soul that needs to be saved) even though I may disagree with behaviors and personal choices. Just because you may go one way to get to a destination and I choose a different route does not mean that we should not be friends. If we still get to the same place at the appropriate time, it will be well. I could potentially miss out on an important relationship if I choose to "tolerate" you and your belief system instead of accepting you in all your glory just the way you are. If I can be fearfully and wonderfully made, so can you.

So, you wanna be my friend?

5 LONG DISTANCE LOVE

Missing you gets easier each day…
By Nyisha Williams

"Missing you gets easier every day because even though I am one day further from the last time I saw you; I am one day closer to the next time I will see you." This quote is a reminder of how the days may seem long in one aspect but short in another. Having the role of a mil-spouse and faced with deployments will test, try, and make you sometimes blue in the face while learning how to express love for your soldier, especially during deployments. It's called "long-distance love." During my journey of long-distance love during deployments, I was faced with really learning how to capture my soldier's attention emotionally and sexually without him physically in my presence. For me, this was the ultimate struggle. How was I supposed to feel love and stay connected to the heart of my husband thousands of miles away from our children and me?

Back in 2008-09 was the very first time that our family would experience an army deployment. My soldier was 23, and I was 22 with a two-year-old son. Need I say more, lol. We were so young, and let's not forget we had only been married for a whopping two months. We were the obvious, premature in almost all of what would be considered important components of life. We had no clue about what was happening and how to handle any situation that would happen. Needless to say, we were "winging it" and "going with the flow." However, Braxton, my husband, was quickly identified as a leader and found himself in the group of the older and most senior soldiers. However, this group of men took care of my soldier and helped to make his first deployment as smooth as possible, all while learning and being developed on becoming a great soldier in the army.

Once in country to Afghanistan, the adjustment phase was the most difficult. Now I am what some would call "married but single," and with a child, after a few months into the deployment and establishing a routine, and learning the time difference, we began adjusting. Even with all these things now going well and me thinking to myself, "I got this," and "you can do it" being my personal motivator, something was missing. I began to realize, I was lying in bed alone, and his feet would not be there to accompany my feet. The smell of his cologne had vanished our bedroom, and importantly, when the moment would make itself known, he was not there to fill the void of being intimate with me. However, the reality of another obstacle that I did not think about in the beginning had set in. I was literally a married but single wife and mother in a long-distance relationship. I had now been introduced to yet another component of deployment and was the least bit prepared.

After exploring all of what was missing physically, emotionally, mentally, and sexually, my husband and I began to do like most, and that is make the best of what you have considering the circumstances. However, we also learned quickly how to be spontaneous. And with that being said, we learned how to use video chat and work around the difference in time zones and schedules that we both had. I began to send letters, and I would spray my perfumes on the letter as well as kiss the letter in one of my red lipstick. Just to spice things up a bit, I would also send in the mail some of my sexiest lingerie sprayed down in perfumes so he would have something in reach with a favorite fragrance he likes for me to wear. However, video chat and lots of pictures of the kids and I was a plus; he enjoys showing off his wife and kids.

Fast forward to years later, and currently, in

deployment status, things are completely different. The times have changed with reliable internet services to stay connected via phone calls and video chat. Mailing to an OCONUS station is quicker, which makes sending spontaneous care packages more exciting when your soldier is not expecting a box. But the one and most important tool that helps to stay connected to my soldier is through our bible plans. We each take turns choosing a bible plan to read together. Doing this allows us both to continue to connect on a spiritual level with keeping God in the center of our union. We also use our topic to check-in for accountability to each other as a reminder not to forget who and what is important in our marriage. We also keep each other involved in our daily routines as if he is home. Choosing to make the sacrifice and compromise in a marriage of a soldier is very important. There are days when you are overwhelmed, and it's too much, but allow yourself to express any emotions you feel and have a support system that will allow you to have a safe space to feel how you feel without judgment. Do not mask or hold your emotions, you will explode at some point, and it may not be the right time for your soldier, which will affect his mission and duty for that day.

Things I would offer as advice to all new and existing military spouses as it relates to long-distance love, keep yourself busy and not get settled too much into a routine. Also, the thing that worked for me was utilizing my soldier's love language. Anytime he sends pictures working out or his new haircut, I compliment how sexy he looked. And tease him about how he needs to stay strong for our adult playtime when he returns home. Speaking his sex language is a plus and keeps him motivated and more to look forward to when he returns home. Keeping him smiling and practicing good communication will give you both a sense of relief that everything is everything, and you both are still in each other's ear. Also, sending video clips of our children when they are playing sports is a must. My soldier enjoys seeing how much the kids are working hard to make him happy, especially when he is not home. One must remember to be aware of your mil spouse and their demeanor. Know when and how to have conversations about things that are happening at home that may not be so good. Timing is everything and will determine the outcome and trajectory of your mil spouse day. If you are a believer in faith, you would want to stay connected and involved in a body of Christ that can encourage you and your family as well as keeping the family covered.

Unfortunately, being married to a soldier, you will face loving your soldier long distance. Treat him or her as if they are at home with you. Practice communication and practice patience; patience will be your best friend. Being in a new environment is tough on your spouse, and it's difficult for them trying to adjust, stay mission focused as well as continue operating in the role of a spouse away from home. Be spontaneous and creative with your gifts and pictures.

Remind him or her daily of how important they are while expressing how grateful you are to be there spouse. Supporting your spouse in a genuine manner will make his or her day stress-free and focused centered. Be his biggest cheerleader and remember to always show your love in the way your soldier receives via his love language.

6 NAVIGATING LOSS TOGETHER

My husband, my rock
By Shawn Jackson

How do you recover from losing your child, your son, your first-born? It's often said that a parent shouldn't have to bury their child, yet that is precisely the gut-wrenching hurdle my husband and I had to face three years into our marriage.

After being told, it would be difficult for me to conceive, by a miracle, I discovered I was pregnant just prior to my husband's four-month-long deployment to Afghanistan. We, of course, were both elated and overjoyed at the unexpected news as any hopeful parents would be. There was just one little hiccup; my husband wouldn't be present to experience the joy of early pregnancy with me. To compensate for this dilemma and the distance between us, I regularly sent pictures of my growing belly and updated him on the changes in my body, appetite, etcetera. Everything was seemingly going according to plan until my eighteen week regularly scheduled check-up appointment. It was then doctors discovered my son had a rare birth defect that, in their opinion, was not sustainable with life—they advised me to terminate the pregnancy.

How could I make that decision alone? How could I choose to end a life that I believed God allowed me to create? More importantly, how was I going to get in contact with my husband, who was on the other side of the world to help me navigate one of the most significant decisions of our married life.

After reaching out to my husband's First Sergeant, the Red Cross was able to locate him, and we finally were able to discuss the situation. I told him what was going on and the time constraints we were under to make our decision. According to the doctors, we had to decide prior to reaching twenty weeks gestation as a termination after that point was unsafe. Within a few days, my husband was back home, and we stood united and made the unpopular decision to continue with the pregnancy and let whatever the consequences of that decision happen naturally. We felt it was not our decision to pick and choose what was a viable child or not; it was God's.

I wish I could tell you everything was smooth sailing after that. On the contrary, that decision was just the beginning of our journey. For the next eighteen weeks, we were peppered with specialists, nay-sayers, advice contrary to our believe system…there were times I would leave the doctor's office in tears wondering how I was going to make it another day. But in all of that, our union as a couple was solidified. We knew we could only depend on one another and God; as a result, we built a protective fortress around our decision and our union.

My son, Aaron Josiah, finally arrived at thirty-six weeks, four weeks premature weighing in at six pounds, one ounce, and sixteen inches. One day post-delivery, he had surgery to correct his defect; however, the damage was irreparable. He was diagnosed with an occipital Encephalocele, a form of a neural tube defect. That meant his brain was growing outside of his head in a sac. As a result, he had malformed brain tissue and only a viable brain stem. Since different areas of the brain control major functions of the body, he came into the world with some significant challenges. He was blind, couldn't regulate his body temperature, had frequent seizures, and eating apnea. We didn't care, we understood all of that, but to us, he was just our precious son.

After a few days, he was released from the hospital, and we took him home, not knowing how many days or even hours we would have him in our lives. Day after day, he kept on surprising everyone around him. He kept on living and, in some cases, thriving. Raising him was not without challenges, but we learned to accommodate his special needs, and he seamlessly integrated into our family and our lives. He was everyone's baby because he was a miracle—the baby that shouldn't have even been born.

Months passed by, and the reality of his disabilities started to unfold slowly. I began to notice he didn't cry to trigger hunger, needing a diaper change, etc. He couldn't maintain his body temperature, so he had to stay on a heating pad constantly. Your brain is what grows your head; as a result of his disability, he had microcephaly, a condition that causes your head to be disproportionately smaller than the rest of your body. Overall, he didn't develop like other babies. There was no crawling, saying "mama", transitioning from formula to solid food, etcetera. In all of this, military life still went on. My husband was again deployed when our son was eleven months old.

Up until this point, though he had some challenges, he didn't have hospital stays. For some unknown reason, he was hospitalized two times in a short period of time with pneumonia while my husband was deployed. During the second hospitalization, I became sad to see him hooked up to all the machines and struggle with the tubes affixed to him. I thought to myself I really don't want my son to have to live a life constantly in and out of the hospital. The doctors again didn't give him much of a good prognosis…it was always touch and go. Like de ja vu, I had to reach out to my husband's unit to have him contacted and returned to the states.

Our son was released from the hospital shortly before his first birthday. We were so excited he again beat the odds we threw him a party at our home to celebrate his miracle year of life with our close friends. Eleven days later, he transitioned to be with the Lord. To say I was hurt and devastated is an understatement. For twelve months, I had stood on my faith and believed our son would be healed, and to have that not been the outcome sent me in a tailspin of emotions. Again, my husband was the rock. He was the steadying force for me to change my perspective of looking at our son dying as a loss to appreciating the time we had with him and the impact he had while he was here. He helped me to focus on the blessing he was to us for the time we had him with us.

Often when a couple goes through something so impactful in their marriage, it either tears them apart or brings them closer together. Our union was strengthened because we choose to cling to one another, our faith, and not allow the circumstance to overwhelm us. How did we do it? Once we made the collective decision to continue with the pregnancy and our reasoning for choosing that option, we had something to refer to when times were challenging.

In the years since our journey with our firstborn son, we had two more "miracle" baby girls and will celebrate twenty years of marriage in September 2020.

7 UNITED FRONT

Home is not a battlefield
By Stephanie Echevarria

As I stand pier-side with my two children in tow, watching my husband's ship set sail for a six-month deployment, I fight back the tears, which threaten to fall, with a fake smile. Standing glued in place, repeating in my head, *stay strong; I can do this. It's only six months, and the deployment will be over before I know it, I* try to muster the strength to deal with the challenge that lies ahead. The ship is well on its way out to sea, and it's time to gather my girls to head home and adjust to a new reality.

Are you familiar with Murphy's Law, the "whatever can go wrong will go wrong" adage? Later that evening, Murphy's Law showed up at my home prepared to do battle. My washer overflowed, the dishwasher stopped working, and my two usually well-behaved daughters seemed to have lost their minds. It felt like my appliances and my children conspired to drive me crazy. I was outnumbered and needed a game plan to get back in control. As each month passed, and new challenges arose, I felt myself changing. I couldn't explain or define the change, but I felt different.

I've survived! It's now a week before the ship's return, and I can see the light on the other side. Excitement is building, and I can now allow myself to think about him returning home. During the five months, three weeks, and eighteen hours, I could not allow myself to focus on how much I missed my husband for fear that it would keep me from holding things together. I didn't think it was possible to miss someone as much as I had missed him. The girls and I planned a big homecoming dinner with all his favorite foods. The kids have been working on a spectacular sign with lots of glitter to hold up for him to see, when the sailors' man the rails as the ship pulls in dockside.

The day has arrived, he is home, and I can breathe a sigh of relief and thank the Lord that he made it home in one piece. The family can return to some sense of normalcy now that he is safe at home. However, normalcy doesn't return because that feeling of change that I felt before is sitting in the corner like a plaid elephant smoking a cigar. The girls miraculously regained their senses and were sweet as pie, flashing big smiles, and acting very politely. That feeling of change deep inside of me was growing stronger with every giggle from my daughters and with them saying, "Things are much better when you are home, daddy." If he only knew how his little darlings were cutting up in his absence, he wouldn't be smiling and grinning. He instead would be issuing punishments and canceling their weekend plans of fun. I held my tongue because I didn't want to be a wet blanket. This reunion isn't going the way I had hoped.

Our loving reunion has been cut short by petty arguments, sniping, and short tempers. I began to wish for his ship to leave again secretly, so the arguing would stop. He has now been home for three weeks, and, just about every night, my husband and I have had some type of disagreement. All the rules we established before he went away for our daughters have been thrown out the window because he feels guilty for being gone. Not once did he ask me if I needed to take a spa day, and, most importantly, there has been no acknowledgment from him on how well I've managed the house. I cannot believe the nerve of that man and his lack of empathy. Doesn't he realize how difficult it was for me juggling all of the craziness that had transpired over the last six months, and that I managed to hold down a full-time job and take grad classes? He has noticed the change in me, but he's not happy with it and says I am pushing him out of our family. He accused me of purposely not making room for him so that I can keep the power. If I am honest with myself, there is a little truth in his words.

How do you keep your home from turning into a battlefield after a deployment? First, start by acknowledging your feelings as valid and not as selfish. Communicating and being open and honest with your spouse is vital for a healthy marriage. The past six months have been trying, and understandably, the experience would have a lasting impact on you in some way. Once everyone has had a chance to decompress, have a conversation about the deployment. Start by asking him how he dealt with the separation and if he had any challenges that he may want to share. Listen carefully and hear what your spouse is telling you. Try to get a sense of what it was like for him being away from their family and home for an extended period of time. You might be under the impression that his time away wasn't stressful or worrisome. Your spouse may not want to bring the drama or stress of his job into the home, so he is less likely to share freely. Sometimes just asking goes a long way in communicating concern. Let him know that you are there as a sounding board, and you want to know about that part of his life. You may be surprised at the response you get. If you want your spouse to listen and hear your concerns, you must be willing to do the same.

Share your experiences with your husband and try to find some humor in the mayhem that you have survived. You can't possibly think he somehow summoned the universe to turn your children and your appliances against you. Let's just chalk that up to bad timing and kids taking advantage of an outnumbered parent. I'll circle back to the children later. Explain to him how overwhelmed you felt during his absence. Acknowledge that his contribution to the smooth running of the family was indeed missed, and you are happy that he has returned. Also, acknowledge that you may have been pushing him out because it's difficult to let go of the reigns completely because you know he will have to leave again. Ask for his patience and understanding as you find your footing. Above all else, make sure that you leave room for him to return home to his family. Make it clear that you're not trying to replace him or eliminate him. He must know you value him as a member of the family. There is nothing worse than feeling rejected by the people that should be welcoming you with open arms. Unfortunately, leaving is a part of his job, and he has no control over that aspect, so it is unfair to punish him. Marriage isn't always 50/50. Sometimes the distribution of stress and responsibility isn't always equal.

Now for those little darlings that turn into hellions when they think an opportunity has presented itself for them to get over. Above all else, you and your spouse must remain united on the rules and discipline structure laid out for your children. Either of you changing rules because of a deployment or any other life event sets a bad precedent. It's best to keep a consistent structure for your family so that everyone knows their roles and what is expected. If there has been some breaking of the rules while one of you was away when the other returns, sit down and come up with a solution together. Let your children know that both of you agree to the punishment that you are handing down. It's only human to want to be the fun-time parent after you have been absent for an extended period of time. You want to spoil your children to make up for the guilt that you feel, which inadvertently undermines the parent who was at home with the children and causes unnecessary strife. Try to resist this pattern and save yourself the unnecessary grief in your marriage.

Once I learned how to take a step back and to see things from my husband's perspective, I realized that I was being unfair. My husband also realized undermining me so that he could win over our daughters was unfair. We reached out to the marriage ministry at our church for support. I was relieved to discover we weren't the only ones having trouble adjusting after long deployments. The important take away is good communication, good listening practices, and empathy for your partner can keep your home from becoming a battlefield full of resentment.

8 OVERCOMING IDENTITY CRISIS

Woman, know thyself
By Robin Major-Oliphant

The Problem

If you had told me ten years ago that I would be a military spouse, married to a United States Airmen at that, I would not have believed nor entertained you. You see, the stigma of military marriages was enough to taint my view on the idea in and of itself. Let me explain, there are a few things you hear about "military men" and military marriages in general, but the one that would've been a deal-breaker for me was the divorced rate in comparison to civilian society. Many people believe that "military men" cheat as soon as they deploy or get orders to move away from their wives. While this can certainly be true, it is the exception and not the rule.

By the time I met my husband, I had been through so many other things in life and was ready to settle down but not initially willing to settle down with him. I had finally gained my footing in life as a single parent, was thriving in a lucrative career, and becoming settled with the idea of South Carolina being home for me. That said, there was no way that I would ever settle down with someone who was younger, unsettled, and furthermore, in the military.

After weeks of avoiding this young man, and then twenty months of dating, we released our inhibitions and said, "I do." I left my six-figure income, friends, home, and safe place, all in the name of love. What I did not realize at the time, though, was that my career was a big part of who I had believed myself to be and leaving it would be more devastating than I initially thought. In retrospect, I always saw myself as the "Robin who sold medical devices," "Robin, the social butterfly," or "Robin, who had nice things", never for a second recognizing that none of these things truly encompassed my identity. They were just things that I did, or possessed, but definitely not WHO I was.

Not knowing who I was would nearly end our marriage before it began because I looked to my husband to fill every single void that he was incapable of doing in his own strength. I was needy, clingy, jealous, and unreasonable. When I moved to Alabama to be with him, he had already established social circles, friendships, and a norm that did not include me. This seemed to come effortlessly for him since it was a part of the military culture. Many of his friends were military, younger, and single, so none of that aligned with my personal interest. Although my husband tried desperately to include me and grappled with ways to make me happy, before I knew it, I was depressed, mean, and angry, sabotaging our marriage before it had even started. I was angry that I had to settle for a mediocre job because several had turned me down, angry at the fact that I struggled to connect with like-minded individuals that I could call "friend", and angry at the fact that he couldn't seem to understand my sheer and utter discontentment.

As I came to the end of myself, paralyzed by the lies of my distorted reality, I knew that I needed to connect with someone who could help me be a better woman, wife, and mother because depression had defeated me in every area of my life. Since I am a PK (preacher's kid), I was almost always prompted to pray and reconnect spiritually when trouble arose, so this time would not be any different. I reconnect with God, this time for relationship in hopes of gaining clarity and relief regarding the toxicity of my life. Reconnecting spiritually, no doubt, was the cornerstone of my overall emotional, mental, and spiritual healing. I also partnered with a faith-based, wife-centered organization to help me walk out the healing and nurturing process gracefully. Spiritual reconnection and divine connection were the perfect recipe for me to overcome the battle that often took place in my mind and manifested in my personal life. Over time of healing and connecting, the revelation of identity-crisis came to me. My depression and self-sabotage were not products of me leaving my career, relocating, and entering into a new marriage, the root of the problem was that my husband had married a woman who was not whole. I had no idea who I was: what made me tick, what did I enjoy doing that was not connected to my husband, children, and my former career, what really defined who I was.

Self-discovery is a beautiful thing. I never considered the effects of leaving my career, safety net, and security because the truth of the matter was, I thought I knew who I was and did not even realize I was allowing circumstances and temporary possessions define me. The beauty in it though, is that marrying a military man and becoming a military wife has pushed me beyond my comfort zone and helped me to discover my true passion, but it didn't happen overnight.

The Process

Clearly, I was focusing on the wrong thing. I was so busy focusing on everything that I sacrificed and everything that my husband was doing but failing to gain awareness, cultivate who I was, and what I was supposed to be doing. Instead of clinging to my husband, forcing relationships with his friends and co-workers, and being envious of his joy, I switched focus completed. Easier said than done, I know, but one thing

I can tell you if it is forced, overbearing, and your intentions in doing it aren't pure, chances are none of the temporary satisfaction you gain from it will last. Either you'll continue to spiral downward, or you'll grip your husband so tight that you'll lose him.

As I let go of resentful thoughts and negative emotions, I began reaching for things that truly brought me personal fulfillment. Let's just call this "getting back to me." I began reading and writing more, connecting with local women, and spending time outside of the house without my husband. This made us both happy. I was smothering the poor guy. The more I gained a true knowing in my spirit of who I was because that was what God said about me, the less I relied on tangible things to bring me pleasure.

Picking up a hobby, going back to school, and establishing a healthy circle of right-minded friends was what brought me temporal joy; however, the true essence of my inner joy and peace came from knowing who I was in Christ. I'm sure some of you may be wondering, "Who is she in Christ, cause' I still don't know". Let me tell you a few things about me: I am a chosen heir, a daughter of the king, I am powerful, I am love, and I was born to co-create in an effort to advance the kingdom of God and fulfill The Great Commission. I know it seems deep or whatever, but everything I have said here either cancels out the negative emotions and thoughts that I struggled with an identity crisis or provide guidance and instruction toward the fulfillment of my assignment.

The thing is, as wives, we do ourselves, our marriages, and husbands a disservice by attempting to walk out wifehood blindly and unaware of our own identity. Knowing and accepting our identity is the cornerstone of being made whole. Marriage does not make you whole. Your husband does not make your whole. Your career, status, and circle of friends do not make you whole. The only one who can make you whole is Christ through awareness and acceptance of what he says about you. We are ever evolving, but the one thing that never changes is our identity in Christ lest we stop believing and accepting the words he spoke over our lives. Trust me, if you want to overcome the trap of losing your identity to your husband, his career, and his needs, discover your identity in Christ. That is where you will truly obtain joy and receive direction for walking in purpose.

9 UNOFFICIAL COMPETITION

Home always wins…
By An Anonymous Military Wife

My story of being a wife, mother, and woman in the military covers eleven years in the Air Force. My entire career to date has been spent dating, loving, marrying, and fighting with and for a life with the same man. Everything in my military experience has been high-speed, and relationships have been no different. Apart from the only world and family I ever knew, I absorbed myself in work and training. I studied hard, worked hard, and played hard. It was in play mode that we met. Like a scene straight out of *Love and Basketball*, we caught a whiff of each other on the basketball courts in beautiful Monterey, California. He captained the Air Force men's team when I started making my name on the women's team. I arrived in January 2009, and by

Valentine's Day, we were exchanging hearts. He left July 2009, for his next training, and that would begin our long-distance relationship. He continued onto Georgia as I completed training in Texas. We had not lived in the same time zone since our initial five months of carefree dating, but in November 2010, he proposed, and rather than suffer a long engagement, we married one week later. I have counted and recounted our time together because, surely, we were dating longer than five months, right? Surprisingly, we had not. In our five months together, we decided that we held significance in each other's lives. In the fourteen months that followed, I moved through three different assignments, created an entirely new life, and grew into a clearer image of who I wanted to be. Our long-distance continued another year through a one-year short tour in Korea, but lucky for us, the Air Force released me of my extended tour selection that would have made it two years. Now came the true start of our life together.

With one year of long-distance marriage under our belt, we left our bases to begin a collocated married life stationed in Maryland. Being a dual-military couple without children meant that we were legally married, but functionally, we had a glorified dating life. We stayed there together for eight months, working different but overlapping shifts until I picked up an opportunity to deploy. I say opportunity because I volunteered. Deployment meant I could do something new, something that could help me stand out. I never worried about the time I would spend away from my husband. I never even asked his thoughts on me leaving. However, I noticed the hesitation and unfamiliarity of others whenever they realized they had made one of two assumptions. The first assumption was that he was the deploying member. The second was that I was a "dependent." Many were hopefully embarrassed and self-chastised by their prejudices, but I also had to admit that, for me, marriage had taken a backseat in the early years. If I am honest, it often still does. We had not built a "normal" married home. In our first six years of marriage, my husband and I spent less than a year in the same home as we chased individual passions. Had I given him his way, we would have become parents a few times over in those years, but I was not ready for competing priorities. I wanted a chance to live. I wanted an opportunity to be seen as the woman I saw in my eyes before what I felt was settling into family life. I just wanted the world and was not convinced that I could have it all.

Working in the same field, my husband and I often unofficially competed against each other. He outranked me, but it wasn't long before I caught up and became the ranking member; yet, I have always felt like I stood in his shadow. Whether for awards, promotion consideration, commissioning, or special assignments, I consistently found myself second to someone I bested in every test and performance board. I told myself that it didn't matter as much because our home still won, but it increasingly mattered to me, as I continually saw myself before shut doors that had just opened for him. I have often questioned how many of these situations were rooted in enduring sexism and prejudice in the military, and how much stemmed from my insatiable pursuit of value through success. My questions have yet to meet definitive answers. In my time, I have received thinly veiled recommendations to separate and follow my husband because we both "didn't have to be in," and I always smiled politely at the well but misguided intentions. I have had my desire to commission discouraged because "what about my family." The fact that my husband has never vented about encountering any of the same frustrations have caused us tensions on more than a few occasions, but each time we managed to find a stalemate that placed the blame on the system and not each other. I wish I could say at this point that I have figured out what it means to be a female veteran, wife, and mother to a daughter, but until I do, I try to learn to be okay with just balancing as best as I can. My journey has taken me through 4 funerals for immediate family members, 3 miscarriages, 6 station changes, 2 deployments, and countless counseling sessions, but it has also brought me 6 promotions, a beautiful daughter, 5 degrees, my first home, and a life of my own designing.

10 OVERCOMING INSECURITY

There is life after infidelity, there is hope after infidelity, and there are love and trust after infidelity
By An Anonymous Military Wife

Where did we go wrong? That was the recurrent thought that caused more unanswered questions to flood my mind as I road silently next to my husband in the taxi headed home from the airport. Overwhelmed by the immediate feeling of insecurity, disappointment, and shame, I allowed to let the silent tears fall. My marriage was over. Just six short months after saying "I DO," I was beginning to believe that my marriage had failed before it had even begun. To make matters worse, I was oblivious to the fact. I honestly did not see this coming.

Just a couple of sobering hours prior to the silent taxi ride, I received an anonymous phone call. I dismissed the call and continued to pack. My husband

Braxton and I were preparing to head back home to Seattle after our weeklong vacation in the Bahamas. We were there in celebration of our long-overdue honeymoon. My husband being enlisted in the Air Force, often always made it impossible to make plans long term, so we often had to take advantage of opportunities like this as they came. We eloped six months prior in June; however, just as we were entering into our nuptials, Braxton was on the heels of a promotion and had to go away for training right after. Therefore, we married on a Friday, and he was in Texas for training by Monday.

The anonymous caller persisted, calling three times more. As the third call was coming in, agitated, I told Braxton that the private caller kept calling back. As he headed out of the hotel room to take our bags down, and get us checked out of the resort, Braxton nonchalantly urged me to answer it. So, I did.

"Hello?"

Caller: "Yea, uh, is this Braxton's wife?"

"Yes, who is this?"

Caller: "So your husband is sleeping with my fiancé' and…"

Click I hung up the phone. This clearly had to be a joke! In hindsight, I'm not sure if I hung up out of denial or if I hung upout of fear of hearing what would come next, but I hung up before even allowing the caller to finish. The caller consistently called back until he realized that I was not going to answer. So, he began texting. For some strange reason, the text messages were easier to take. They didn't lessen the blow, but it gave me time to process at least for a moment as the information flooded my phone. The more information that the man in the text messages provided about my husband's alleged affair, the more I pushed back with questions demanding proof and sinking further into denial. By the time he provided proof in the form of screenshots from his fiancé's phone, which included sexual expletive and more, Braxton had re-entered the hotel room. I didn't question him right away; I needed more time—time to process it all.

I took the entire six-hour flight from Nassau to Seattle to "process." By the time we made it into the taxi in Seattle, I had allowed my emotions to rip me to pieces. Seeing the silent tears falls from my eyes confirmed my husband's suspicion that something was wrong. He asked a few times, and I just told him that I was worn out from the trip. I'm sure he knew better but, I didn't care at the time. Finally, when he asked as we pulled up to our condo, I said, "Yes something is wrong. I know about your affair." If the pain from having to mutter those words weren't enough, his response in comparison to my expectations was not making the situation better. He didn't deny it, as I expected him to. Instead, he questioned me about why I didn't say anything earlier, as if I had done something wrong. Maybe this was his way of "processing."

The days to follow were rough. I had not struggled with insecurity since I was in my twenties. There was something about infidelity that took me there. Prior to infidelity, I had been the most confident, bold, and fearless person I knew, and if you asked me, I was pretty sure of everything going on in my life. Despite the often-trivial situation that military life provided, I was sure, and I was secure- until now. I thought that my marriage was solid. I thought that I was a good woman, a good wife, I thought that I was enough, but for me the infidelity of my husband took all of those feelings away from me, and replaced them with doubt, insecurity, and hopelessness.

God, your God, will restore everything you lost; he'll have compassion on you; he'll come back and pick up the pieces from all the places where you were scattered. Deuteronomy 30:3 MSG

I reached out to a trusted sister friend and told her about the affair. She listened, not passing any judgment or giving any "if it were me" advice. She simply asked me two questions, "what do you want to do, and did you ask God?". I became completely undone. I screamed, I cried and ultimately, I told her that I wanted a divorce. The louder I screamed that I wanted a divorce; through my emotional tear-filled emotional escapade, the louder I heard the holy spirit say not yet. This frustrated me to my core. I wanted what I wanted, and I wanted it when I wanted it. I did not care to go through the process.

Oftentimes we prefer to make swift decisions based on our emotions hoping that it eases the pain when ultimately it only intensifies the pain and magnifies the problem at hand. Sometimes we must let a situation breathe.

I realized that in my husband's indiscretions that I had insecurities that I needed to be healed. So, with mixed emotions, I still prayed. Divorce or not, I did not want to deal with the pain of heartache for any length of time. I had done it too many times before, in past failed relationships, and it seemed like it was a pain that I could not escape. I asked God to heal my heart COMPLETELY. Not just from Braxton's infidelity, but from the infidelity of the other men that I had been with in the past. I asked God to heal my broken heart and allow me to forgive my husband first. I did not immediately want to love my husband, unconditionally. I definitely did not want to trust him at all, but I knew that I had to in order to heal and move forward. I also asked God to help me with forgiving everyone else who had broken my heart this way. I was completely wrecked, and I knew it. I needed complete restoration of my heart.

Moving Forward

The Lord spoke to me through Isaiah 26:3, "You will keep in perfect peace those whose minds are steadfast, because they trust in you." It was then that I knew that I needed to take my eyes off my husband and focus on God so that I could heal. Although my husband was the one who had committed the sin against our covenant, it was not up to me to babysit him nor condemn him. My only job was to heal and focus on loving him unconditionally.

I know, I know. It sounds much easier than it actually was to do, however, with Christ, and focusing on Him consistently, the peace that I possessed as I walked the journey of healing from the betrayal of infidelity was indescribable. I am sure that many of you may question "Has he ever cheated again?" "Do you think he will cheat again?" "How did you hold him accountable?".

My answer is simple and maybe political or maybe like somewhat of a paradox, but the truth is whether he has cheated again or not is beyond me, however, I do trust that he has not. I do not think he will cheat again because I trust God to sustain my marriage and to give him the temperance and restraint he needs if he ever gets tempted. As far as holding him accountable, that isn't my job. The most I can do is love him, treat him like a man and not a child, and cover him in prayer. I am happy to say now ten years later, our marriage is joyful, blissful and that we keep God at the center.

In infidelity, I realized that there would be several unanswered questions. The more we try to answer questions and control the narrative, the more we prolong the process of healing and growing through the affair. There is life after infidelity, there is hope after infidelity, and there is love and trust after infidelity. You must be willing to focus on God and trust Him to renew your covenant.

11 BETRAYED BUT I STAYED

Rebuilding is intentional; rebuilding takes time
By Monica McGinnis

What I encountered was a run-in with infidelity in my marriage of sixteen years. Instantly I was consumed with internal questions like, is it my fault or was told about the infidelity when it first happened. My reality was that my intuition had been confirmed because I'd suspected something was going on but chose denial- it was easier. Of course, this was my immediate emotional response to my feelings and fleshly thoughts.

I found out about the infidelity in my marriage while my husband was on a Geo Bachelor Military Duty tour, away from the family for three years. Maybe we opened the door and allowed infidelity to infiltrate our marriage, or maybe it was to grow us and make us stronger Christians who strive to be more like Christ daily. Yeah, I'll be honest. I didn't think of the spiritual at first when I was presented with this major life-altering issue it in hindsight; I believe it to be the latter.

Dealing with infidelity requires a support system. Regardless of whether you make the decision to stay or go. It is a traumatic experience that you must recover from individually; and collectively if you decide to stay. Once I realized I needed some support, I called on some ladies who have been a great support in my life, who also knew my husband and myself very well. I refer to them as my prayer warriors. From the moment I contacted the prayer warriors, I received words of encouragement and shared sentiments such as "I've been there too." They assured me that rebuilding from the place of pain I was in takes time, so that was the path my husband and I chose to walk together.

I made the decision to re-establish my communication with the center of our relationship, our Lord and Savior Jesus Christ. I connected daily in my prayer room, crying at times and praying for God to do a work in myself and my husband. Of course, this first step wasn't easy because I wanted to understand why. Why had my husband decided to step out on our marriage on top of us being separated by duty, why weren't we basking in the blessings God had already given us? The truth of the matter is, we as humans fall in multiple ways, and Satan has a way of appealing to our flesh if we aren't built up in our armor spiritually. I wanted to wallow in my pity at times. I was fully broken down; I didn't want even to leave the house.

Pain is such a strong emotion that when we get there it cuts us deep and bleeds a lot. When we know we are in pain, it is easy for us to avoid others and not want to be around others. Isolation—it feels good, almost comforting in our pain but is the enemy's territory to keep us there and manipulate our thoughts. I thank God for my sisters, who knew what I was going through. Daily, they helped me through contact and prayers, helping me find my way back from the dark place of pain. This was most beneficial to me and my marriage to keep intact and get even stronger because there was no way that we could heal and move forward as long as I was stuck in the place of pain.

Give him your mess; give him back the blessings. God goes where he knows your faith will multiply.

In hindsight, I realized I placed my husband on a pedestal; yes, way up there. Meanwhile, not giving myself much of any credit even though I was doing everything I needed in my marriage, as a wife and mother. I felt as though my covering was broken. It was time for me to give my marriage back to God. Broken was me in the hands of Jesus multiplied by God. We needed to be broken; we needed to grow. We were living a fairytale, and we needed to learn a deeper lesson. I didn't see it at the time, the lesson that is. I did, however, keep asking God why me, as I grappled for answers. I listened, and He spoke. God's instruction was for me to get a deeper understanding of my relationship with him. Honestly, I wasn't ready to hear that because this situation made me look back at all the things I had ever done and how I treated my husband and, most of all, where I placed him in my life. To go further in marriage, we must have Christ as our center, and I believe he takes everyone there.

My marriage was going through the fiery furnace, and it hurt a lot, I wanted to give up lots of times. My husband still being away did not make it easier. I often emailed for comfort, reassurance, and answers, but it was all temporal. I fell into a clinical depression and even went to the doctor for prescription depression pills. I prayed more and more. Yet, I was still very depressed. In spite of my depression, God had me covered and giving me the patience to stay in the fight. That was a lesson I was learning and continue to learn.

No, I was not the perfect wife at the time; truth be told, I'm still not. I have done things to upset and make my husband not want to be around me; I'm sure I still do, but the fact of the matter is that we had to learn to communicate our feelings with one another regardless. We needed to submit ourselves, our marriage, and our needs to God. He is the only one able to keep us, sustain us, and provide what we needed—

when we needed it.

We went through months of trying to reconnect and many months of counseling. Years later, we are still learning to connect. We decided to stay in our marriage and fight and surrender ourselves to God. When our faith is tested, we're going to go through. We must prepare ourselves to stay in the ring and be tried. I am not saying any of this came easy because it didn't. I wanted to put on my big girl panties, woman up and leave because no one wants to be betrayed especially by their spouse No one desires to be in a relationship where you are in a home with someone you don't feel you trust, and no one wants to be touched by someone who they once held so highly, but now out of obligation the touch they allow feels dirty.

The way we feel is reflected in our thoughts and actions. We must have a safe-haven and a comfort zone. Our minds wander, and we must keep them under the subjection of God's word. Not one emotional battle we go through in life is easy, but with God, it is bearable. We must make the decision to run to God and away from the world's way of dealing with issues in our lives. Nothing we encounter in life is new to our God. All in all, this situation in my life made me look back at LOVE. Love is not rude and unworthy and doesn't cause strife.

Consider it pure joy, my brothers and sisters, whenever you face trials of many kinds, because you know that the testing of your faith produces perseverance. Let perseverance finish its work so that you may be mature and complete, not lacking anything. If any of you lacks wisdom, you should ask God, who gives generously to all without finding fault, and it will be given to you. James 1:2-5 NIV

KEEPING UP WITH THE CO-AUTHORS

Misha Taylor
Marine Corps (wife)
Camp Pendleton, CA
www.convosformoms.com

Leah Sledge
Air Force (wife)
Misawa, Japan
www.movethatmtn.org

Stephanie Echevarria
Navy (wife)
Yokosuka, Japan
www.365perfectbalance.com

Melissa McGill
Navy (wife)
Yokosuka, Japan
www.amazon.com/author/melissamcgill

Nyisha Williams
Army National Guard (wife)
Deployed
Connect with me on Facebook @ Nyisha Williams

Join social groups such as Wife Talk Military, The Military Wife, and Black Military Wives on Facebook to keep the conversation going.

ABOUT THE AUTHOR

Robin's heart for God and his people is what sets her apart and demonstrates God's grace in her life as an author, publisher, speaker, mentor, and servant leader. She uses the transparent, tumultuous testimonies of her life as a catalyst for change; to empower others to heal, be delivered from, and overcome the traumas of the past. Robin has been resilient in overcoming real-life situations that men and women face daily such as, domestic violence, sexual abuse, self-esteem issues, and depression. Robin's mission is to empower others to pursue their passion, overcome, and thrive by utilizing her gifts and helping people to discover their own. Read more at www.one2mpower.com

BOOKS FROM THE AUTHOR

From Pieces to Peace: Damaged Goods

A story filled with brokenness, abuse, heartache, betrayal, and resentment. Damaged Goods is a captivating read that grips you from beginning to end. It's the riveting real-life story of Robin Major-Oliphant, who invites you to share in her journey toward peace and wholeness, filled with emotion, and passion.

From Pieces to Peace: Forgiving the Unforgivable

A captivating journey of overcoming heartache, bitterness, and betrayal through forgiveness. The sequel to Damaged Goods; Forgiving the Unforgivable is a story of resilience that provides a glimpse of how Robin becomes an OVERCOMER after living life as "damaged goods".

Freedom Through Forgiveness: An Interactive Prayer Journal and Devotional

Forgiveness can seem like such a daunting task; however, this interactive prayer journal and devotional contains daily scriptures, prayer-prompts, and words of encouragement; from the author that will shift your perspective regarding forgiveness, and the way you see yourself.

Love You: Unconditionally *an anthology*

Have you ever thought, "I wish I knew then what I know now"? Or maybe even wondered what life would be like if you trusted your intuition and followed your dreams. For many, the thought lingers throughout the course of their lives while they remain paralyzed by fear and bound by insecurities. These testimonies include life lessons, self-love, and encouragement for teen girls and young women everywhere. Be prepared to be uplifted, empowered, and inspired to love you, God's way.

Made in the USA
Monee, IL
08 November 2026

67388653R00052